SPOTLIGHT ON NATURE

MANATEE

LISA M. BOLT SIMONS

CREATIVE EDUCATION · CREATIVE PAPERBACKS

Published by Creative Education and Creative Paperbacks
P.O. Box 227, Mankato, Minnesota 56002
Creative Education and Creative Paperbacks
are imprints of The Creative Company
www.thecreativecompany.us

Design by Blue Design, Inc.
Art direction by Graham Morgan
Edited by Jill Kalz

Photographs by Alamy Stock Photo/David Fleetham, 16, Jeff Mondragon, 27, WaterFrame_mus, 21; flickr/Biodiversity Heritage Library, 3, 8, 10, 14, 16, 20, 22, 28; Getty Images/Colors and shapes of underwater world, 4–5, Gregory Sweeney, 12, 14, James R.D. Scott, 21, Jeff Foott, 23; iStock/Amanda Cotton, 18; National Geographic Creative/Brian Skerry, 28; Pexels/Jakub Pabis, 15, Rene Ferrer, 17; Shutterstock/24, Dennis Tokarzewski, 29, Liquid Productions, LLC, 29; Unsplash/Maegan Luckiesh, cover, 1, NOAA, 6, Robin Teng, 9, 10; Wikimedia Commons/Gerson Barreiros, 29, Vardhanjp, 11

Library of Congress Cataloging-in-Publication Data
Names: Simons, Lisa M. Bolt, 1969- author.
Title: Manatee / by Lisa M. Bolt Simons.
Description: Mankato, Minnesota : Creative Education, Creative Paper backs, [2026] | Series: Spotlight on nature | Includes bibliographical references and index. | Audience: Ages 10-13 | Audience: Grades 4-6 | Summary: "Barrel-roll into the manatee's ocean world with this nature title for middle-grade wildlife lovers. Informational text pairs with a narrative about a single "sea cow" family to spotlight the wild marine animal's life cycle, supported by infographics and a timeline of developmental milestones"— Provided by publisher.
Identifiers: LCCN 2024048407 (print) | LCCN 2024048408 (ebook) | ISBN 9798889896258 (library binding) | ISBN 9781682777916 (paper back) | ISBN 9798889897057 (ebook)
Subjects: LCSH: Manatees—Juvenile literature
Classification: LCC QL737.S63 S54 2026 (print) | LCC QL737.S63 (ebook) | DDC 599.55—dc23/eng/20241207
LC record available at https://lccn.loc.gov/2024048407
LC ebook record available at https://lccn.loc.gov/2024048408

Printed in India

CONTENTS

MEET THE FAMILY

FLORIDA MANATEES of Crystal River National Wildlife Refuge

One hour north of Tampa, Florida, inland from the Gulf of Mexico, lies Kings Bay. In and around this bay are 20 islands, bays, and rivers that make up the Crystal River National Wildlife **Refuge**. Established in 1983, this 80-acre (32-hectare) refuge is the only area that specifically protects the Florida manatee.

More than 70 springs pump 600 million gallons (2.3 billion liters) of fresh water per day into the saltwater bay. Otters, black bears, alligators, and hundreds of bird species share the refuge. So do large marine mammals called manatees.

It's April, and Florida's gulf water is starting to warm with the coming summer. Pairs of manatee mothers and calves will soon swim these coastal waters.

A female manatee has waited about a year for this moment. She floats high in the water and breathes every 30 seconds. She rolls over and over. She is preparing to give birth.

CLOSE-UP

Mammary glands

A baby manatee drinks milk from its mother right after birth. Like all mammals, female manatees have mammary glands. The manatee's glands are on the backside of the flippers.

CHAPTER ONE

LIFE BEGINS

Manatees, also known as sea cows, are graceful marine mammals, despite their size. Adults average 10 feet (3 meters) in length and weigh between 800 and 1,200 pounds (363–544 kilograms). Manatees have long, thick bodies that narrow into a paddle-shaped tail. They have two flippers. Their snouts have whiskers. Manatees are gray but sometimes have brown or green **algae** growing on their hairy skin. These animals never leave the water, but they must regularly rise to the surface to breathe air.

There are three species of manatee: the West Indian, the Amazonian, and the African. All are **threatened**. The three species of manatee are generally determined by where they live. West Indian manatees are found along the North American coast from Florida to Brazil. They live in salt water and fresh water. The Amazonian manatee lives in the Amazon River. This species can survive only in fresh water. The African manatee lives in inland rivers and channels along the coast of Africa. Scientists know the least about this species.

MANATEE MILESTONES

- Born in water
- Takes first breath
- Weight: between 60 and 70 pounds (27–32 kg)

Manatees eat a lot. Mostly plant-eaters, or herbivores, they consume up to 10 percent of their body weight in 24 hours. Florida manatees, a subspecies of the West Indian, eat more than 60 kinds of plants. The West African manatee eats entire fields of planted rice. Antillean manatees, another subspecies of the West Indian, have been observed eating meat, namely clams and fish. During dry seasons, when water levels drop and vegetation is scarce, Amazonian manatees **fast**. They can fast for months.

CLOSE-UP

Bones

Human rib bones have a spongy substance inside called marrow. Manatee rib bones are solid. The solid bones make the manatee's body density equal to the water, creating "neutral buoyancy." The manatee doesn't float or sink.

Large groups of manatees are often seen together. But the groups vary and change when it comes to the number of males, females, and ages. Manatees are not aggressive or territorial as they eat, travel, explore, and socialize. The basic social unit of the manatee is the mother and her offspring.

Welcome to the World

Tucked away in a shallow, quiet cove, the female manatee rolls over and over, her tail flexing. As she turns in the water, a baby manatee starts to leave her body, headfirst. Blood that helped nourish the newborn before birth also comes out. As soon as the baby manatee is all the way out, the mother nudges it to the water's surface. The calf takes its first breath of air. For the next hour, it relies on its mother to stay afloat.

Fully grown, healthy manatees don't have any natural predators. Humans, with their boats and fishing nets, are the biggest threat. Sharks, crocodiles, and alligators may prey on young or sick manatees, but they usually don't share the same waters. Male manatees don't protect their young, called calves, but mothers do. They use their bodies as a protective shield or simply swim away from danger with their calf. An orphaned calf cannot survive without its mother.

1 WEEK

- Learns to swim
- Nurses underwater
- Vocalizes with mother

CLOSE-UP

Lungs

Manatees replace 90 percent of the air in their lungs with one breath. By comparison, when humans take a breath, they replace only 10 percent. Manatees can stay underwater for up to 20 minutes before needing to breathe.

FEATURED  FAMILY

First Meal

The manatee calf cannot eat vegetation right after birth. It must nurse from its mother. After its first breath of air, the calf swims close to its mother's body. It knows just where to find one of her teats, right on the backside of a flipper. The calf drinks the warm milk for two to three minutes at a time. Within a few hours, the calf and its mother settle into the cove. The newborn nurses every 45 to 60 minutes as it gets used to its new home.

The **DUGONG,** with its **dolphin-like** tail, is a stocky cousin of the manatee.

3 WEEKS

- Is still nursing
- Starts eating plants
- Starts learning travel routes

CLOSE-UP
Ears

Even though the manatee has no external tissue, it does have ears! There is a small hole on each side of the head, right behind the eyes. Manatee ears are especially good at picking up high-pitched sounds.

CHAPTER TWO

EARLY ADVENTURES

As the manatee calf continues to grow, it stays close to its mother. Instead of swimming single file, like it will as an adult, the calf swims alongside its mother. The calf can swim much easier just behind its mother's flipper. It's also believed that the pair can communicate best when side-by-side.

At birth or soon after, the calf vocalizes with its mother. Manatees chirp, whistle, squeak, and grunt. They make these sounds in fear, anger, or as a simple, "Where are you?" contact, depending on the situation. Since manatees and their young spend so much time with each other, vocalizing is a meaningful bonding experience for the pair.

The mother teaches its calf about **migration** routes, where to feed and rest, and the locations of the warm water shelters. During the summer, Florida manatees have been seen as far west as Texas and as far north as Massachusetts, but it's uncommon. Instead, they travel mostly around

8 MONTHS

- Drinks more milk than ever
- Adds more plants to its diet

CLOSE-UP

Lips

The manatee's closest land relative is the elephant. Like a much shortened, smaller trunk, the manatee's upper lips move separately from each other. This body feature helps the animal grab and pull vegetation into its mouth.

FEATURED  FAMILY

Just a Taste

The newborn calf continues to nurse as it follows its mother through the bays and around the islands from the refuge to the coast of Florida. As they find shallow coves, bays, and harbors, they stop so the mother can eat. As the weeks pass, the calf also starts trying to eat the weeds, algae, and grasses found in the salt water. The mother spends about six to eight hours a day eating. The calf grazes a bit but mostly nurses.

the state of Florida. They're also spotted occasionally around Alabama, Georgia, and South Carolina.

When Florida manatees head back to their preferred mating and gathering habitats for the winter, some have found shelter at power plants. Where the power plant releases water into the ocean, called the **effluent**, it's warm and inviting for the mammals. Conservationists worry, however, that aging power plants will be shut down to make way for new energy sources. If that happens, manatees will have fewer spots to warm their bodies.

1 YEAR

- Can live without its mother's milk if necessary

CLOSE-UP

Body hair

The fine hairs all over a manatee's body help with the mammal's sense of touch. If there are changes in water currents, manatees feel them. This is useful for detecting other animals or boats.

Give It a Try

It's June, and the calf is two months old. It's grown, thanks to its mother's milk. The pair continues to migrate south along the coast of Florida, the calf still swimming at its mother's side. In a quiet cove, the two find a bed of turtle grass. The mother begins to graze, pulling on the thick, inches-high vegetation with her lips. The calf does, too. It takes some practice, though, and after 45 minutes, the calf tires, swims closer to its mother, and nurses. It will try grazing again tomorrow.

ADULT MANATEES

are not afraid of **alligators** and will simply nudge them of out of the way.

(18) MONTHS

- Weans slowly off milk
- Eats mostly plants

CLOSE-UP

Flippers

Manatees have two front flippers. They use them to swim, as well as to touch and comfort each other. The bones in the flipper are jointed and look like a human hand. These joints help manatees hold objects.

CHAPTER THREE

LIFE LESSONS

When the calf stops nursing at about two years old, it leaves its mother. Females will eventually be ready to have their own newborn calves. They are ready to mate between three and five years old. Males are ready between five and seven years old. When it's time to breed, one female, called a cow, is followed by a dozen or more males, called bulls. This group is called a mating herd. The female turns and twists as the males try to stay close to her. The males compete to be the first to mate.

The female manatee's pregnancy lasts about one year. She gives birth to one calf in the spring or early summer. Twins are rare. She separates herself from the herd before she gives birth. The male has nothing to do with his family. Manatees give birth every two to five years.

The mother nurses her calf for up to two years. Once the calf has stopped nursing and is eating only vegetation, it goes off on its own.

Manatees are partly social. When not part of a mother-calf pairing, they may live alone or in a group. In the company of other manatees, they

2 YEARS

- Stops nursing
- Leaves mother

have been observed playing together. Sometimes they bodysurf, riding the currents below flood dams. Other times, manatees play Follow the Leader. Two or more manatees swim in a single file. They coordinate diving, breathing, and changing their direction. Manatee groups are always on the move and always changing.

Manatees are built for underwater life. Their lungs hold air for 20 minutes when the mammals rest. With both rod and cone cells in their **retinas**, manatees can see well in both clear and murky water. Their nostrils seal shut when they are underwater, and their kidneys regulate their salt intake. Calves learn how to navigate the waters from their mothers.

This Is How It's Done

The calf and its mother have returned to Crystal River National Wildlife Refuge for the winter. The calf is now eight months old. Together, the pair swam around Florida, up to Georgia, and back. Not only is the water here in the refuge delightfully warm, it's full of aquatic vegetation, such as eelgrass and hydrilla. The calf is drinking more milk now than it ever has, and it's also gotten a lot better at eating plants. It's been watching its mother carefully. She's been a good teacher.

CLOSE-UP

Teeth

Because manatees eat so often, new teeth replace the old teeth their whole lives. The teeth are called "marching molars." When the front set of molars wears down, the set of molars behind moves, or "marches," into that front place.

3 YEARS

- Females ready to reproduce

5 YEARS

- Males ready to reproduce

Manatees live about 40 years in the wild and 60 or more years in captivity. Although adults have no natural predators, all three species of manatee are considered vulnerable to **extinction**. Human interference, loss of habitat, being crushed in canal gates, and weather-related events threaten the animals' survival.

CLOSE-UP

Vertebrae

Most mammals have seven small bones in their necks called vertebrae. Manatees have only six. One fewer bone means manatees cannot turn their head. They must turn or roll their whole body to see, which makes them appear more graceful.

Practice Makes Perfect

At one and a half years old, the manatee calf is drinking the least amount of milk in its life. It still nurses, but now its intake of vegetation is the highest it's ever been. The calf has traveled with its mother twice around Florida. At each cove and harbor, it has consumed algae and underwater grasses, nearing 100 pounds (45 kg) of food every day, just like its mother. Soon, the calf will stop nursing and leave its mother's side, finding its own way to food and warm waters.

Graceful manatees are the

"MERMAIDS"

of the rivers and seas.

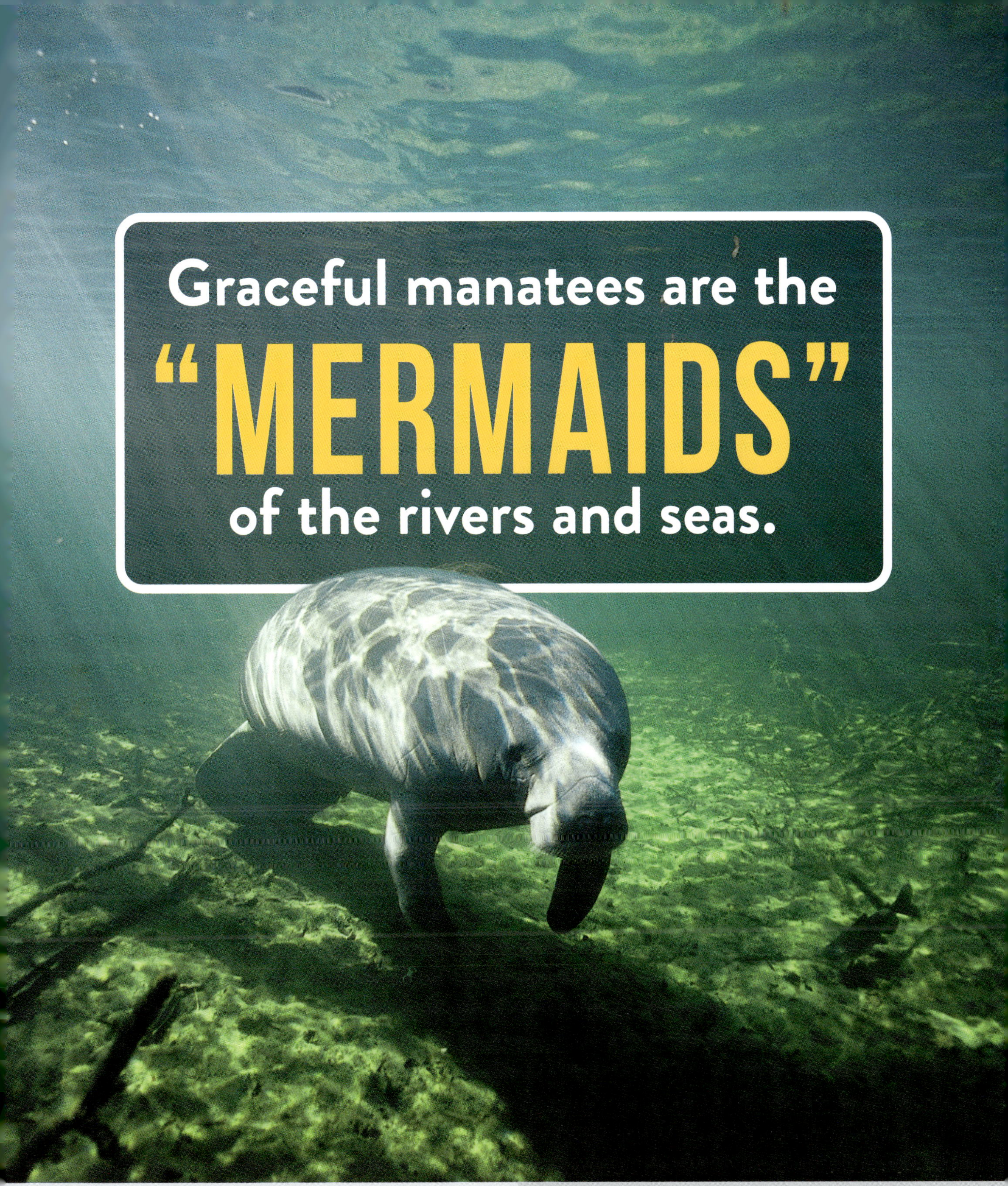

40 YEARS

- End of life in the wild

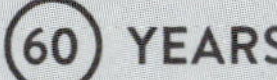

60 YEARS

- End of life in captivity

CHAPTER FOUR

PROTECTING THE MANATEE

World explorers in the late 1400s recorded the first sightings of manatees, claiming they were mermaids. The animal order Sirenia comes from the Greek mythical legend of Sirens, human-like beings that sang to sailors to bait them into crashing their ships on rocky shores.

One of the manatee's relatives was the Steller's sea cow. It grew up to 30 feet (9.1 m) long and weighed close to 9,000 pounds (4,082 kg). First described by explorers in the mid-1700s, the sea cow was hunted to extinction a mere three decades later. It was prized for its meat and blubber.

Today's manatees are often hunted for their meat, too. Sometimes fishing nets catch them, preventing the animals from rising to the surface to breathe and causing them to drown. Boating accidents have caused half the deaths of West Indian manatees. Even if manatees escape death, sharp boat propellers often scar them.

Pollution is also a problem for manatees. Large growths of algae called "blooms" may be the result of fertilizer runoff from farms and lawns. In some cases, the overfed algae releases toxins that can negatively affect marine life, birds, and people. The bloom often turns the water red. Manatees that come in contact with a harmful "red tide" may become paralyzed and suffocate.

But all is not lost. The U.S. federal government is one hope. The Marine Mammal Protection Act of 1972 and the Endangered Species Act of 1973 were both passed by lawmakers to help protect the manatee. As part of these protections, trying to feed or touch the animals is against the law.

Organizations are also trying to save these water creatures. The mission of the Save the Manatee Club is to not only to protect the animal but to also protect its habitat for generations to come. The Manatee Rescue & Rehabilitation Partnership, established in 2001, works with zoos, aquariums, and other organizations to rescue, rehabilitate, and release manatees. The partners also study manatee biology and health and provide information about the animal through public education.

Manatees are amazing, gentle creatures. Helping them continue to recover from their decreasing numbers and protecting their habitat will enable manatees to thrive for generations to come.

FAMILY ALBUM

SNAPSHOTS

The manatee is closely related to two very different animals living today: the elephant and the small hyrax (rock rabbit).

Manatees leave "footprints." When they swim, manatees make oval-shaped ripples in the water that show where they have been.

Although usually slow-moving, manatees can swim up to 15 miles (24 kilometers) per hour in short bursts.

The West Indian and Amazonian manatees have three to four "fingernails" on each front flipper.

Scientists can estimate the age of a manatee by looking at the growth layers in its ear bones.

Manatees cannot be in water that's colder than 68 degrees Fahrenheit (20 degrees Celsius) for a long time. They prefer water that stays about 72 °F (22 °C).

In the 1970s, there were only a few hundred Florida manatees in the wild. Thanks to conservation efforts, there are more than 6,000 today.

Manatee brains are smooth, and the size of their brain compared to their body size is the lowest of any mammal.

The Amazonian manatee must live in and eat in fresh water.

The Amazonian manatee is the smallest of the species. It usually has white patches on its belly.

WORDS to Know

algae a plantlike organism that grows mostly in water

effluent industrial waste material, such as smoke or water, released into the environment

extinction the state of no longer existing

fast to not eat

migration the act of traveling from one place to another, usually for feeding or breeding purposes

refuge a place that provides protection

retina the sensory membrane that lines the eye

threatened not endangered but still in need of protection

LEARN MORE

Books

Feldstein, Stephanie. *Save Ocean Life*. Ann Arbor, Mich.: Cherry Lake Publishing, 2024.

Gish, Melissa. *Manatees*. Mankato, Minn.: Creative Education and Creative Paperbacks, 2024.

Martin, Claudia. *Marine Ecosystems*. Minneapolis: Bearport Publishing Company, 2025.

Websites

"Manatee." San Diego Zoo Wildlife Alliance, Animals & Plants. https://animals.sandiegozoo.org/animals/manatee

"Manatees." National Geographic. https://www.nationalgeographic.com/animals/mammals/facts/manatees

"Manatee Webcams." Save the Manatee. https://savethemanatee.org/manatees/webcams/

Documentaries

Bird, Jonathan, dir. *Endangered Mermaids: The Manatees of Florida*. Tappan, N.Y.: Janson Media, 2007.

Manatees on the Move. Hamburg, Germany: DocLights, 2022.

Nixon, Robert, and Fisher Stevens, dirs. *Mission Blue*. New York: Insurgent Media, 2014.

Note: Every effort has been made to ensure that any websites listed above were active at the time of publication. However, because of the nature of the Internet, it is impossible to guarantee that these sites will remain active indefinitely or that their contents will not be altered.

Visit

CINCINNATI ZOO

Manatees live here temporarily as they rehabilitate before being returned to Florida waters.

3400 Vine Street

Cincinnati, OH 45220

CRYSTAL RIVER NATIONAL WILDLIFE REFUGE

This refuge is the only area that exclusively protects the Florida manatee.

1502 SE Kings Bay Drive

Crystal River, FL 34429-4661

HOMOSASSA SPRINGS STATE WILDLIFE PARK

Walk in the Underwater Observatory to watch the manatees swim.

4150 S. Suncoast Blvd.

Homosassa, FL 34446

INDEX